FLORA OF TROPICAL EAST AFRICA

PEDALIACEAE

E. A. Bruce

Annual or perennial herbs, rarely shrubs or small trees, covered with mucilage-glands (at least on the young parts), which become slimy when wetted. Leaves opposite or the upper alternate, usually simple, exstipulate. Flowers usually solitary in the leaf axils, rarely in few-flowered racemes ; pedicels generally with nectarial glands at the base. Flowers hermaphrodite, irregular. Calyx 5-partite. Corolla gamopetalous ; tube usually obliquely campanulate, sometimes funnel-shaped or cylindrical and gibbous or spurred at the base ; limb sub-bilabiate or subequally 5-lobed. Stamens 4, didynamous (fifth stamen often represented by a staminode) inserted near the base and normally included in the tube. Anther-cells 2, parallel or divaricate, opening lengthwise, connective usually gland-tipped. Disc hypogynous, fleshy, generally conspicuous, often asymmetrical. Ovary superior, usually 2-celled, cells often completely or partially divided by false septa, each compartment containing 1–∞ ovules attached to the central placenta ; style filiform, exceeding the anthers, stigma usually 2-lobed, rarely 4-lobed or trumpet-shaped. Fruit very variable, dehiscent, or indehiscent, often provided with horns, spines or wings. Seeds 1–∞ in each compartment, sometimes winged ; testa often reticulate or pitted, endosperm very thin.

Spiny shrubs or small trees 2–6 m. high; leaves fasciculate on short branches ; flowers in few-flowered racemes ; corolla large with a long narrowly funnel-shaped or cylindrical tube, spurred or gibbous at the base ; fruit a woody capsule ; seeds large, compressed, smooth, winged . . 1. **Sesamothamnus**

Annual or perennial herbs (rarely shrublets up to 1·3 m. high) ; leaves not fasciculate ; flowers axillary ; corolla campanulate or funnel-shaped, never elongate and spurred ; fruits variable ; seeds rarely winged, if so pitted not smooth :

Ovary and fruit 2-celled ; cells undivided ; fruit indehiscent ; corolla-tube funnel-shaped or sub-cylindrical ; anther-cells kidney-shaped, divergent :

Annual herbs ; fruits subpyramidal, 4-angled with a spine at each basal angle . . . . 2. **Pedalium**

Perennial herbs with swollen stem-base and tuberous root ; fruits 4-winged or 4-angled, without basal spines . . . . . 3. **Pterodiscus**

Ovary and fruit 2- (rarely 4-) celled, cells divided by a false septum ; fruit dehiscent or indehiscent ; corolla-tube obliquely campanulate ; anther-cells oblong, parallel :

Fruit woody, indehiscent ; ovary and fruit completely divided by a false septum ; compartments 1–3-seeded :
Fruit disc-shaped with a raised, central, 2-
horned portion   .   .   .   .   . 4. **Dicerocaryum**
Fruit ovoid to subglobose, densely covered with
small spines   .   .   .   .   .   . 5. **Josephinia**
Fruit a many-seeded, dehiscent, incompletely
septate capsule :
Capsule rounded or truncate at the apex and
horned at the angles (horns rarely sup-
pressed)   .   .   .   .   .   . 6. **Ceratotheca**
Capsule acute with a single terminal beak   . 7. **Sesamum**

1. SESAMOTHAMNUS

Welw. in Trans. Linn. Soc. 27 : 49, t. 18 (1869) ; E. A. Bruce in K.B. 1953 :
417 (1953)

Sigmatosiphon Engl. in E. J. 19 : 150 (1894)

Branched spiny shrubs or small trees ; trunks smooth, usually swollen at
the base ; leafless when flowering ; branches ascending, spiny. Leaves
deciduous, usually obovate, fasciculate on short branches in the axils of the
spines (modified petioles). Flowers large, white, pink or yellow, often sweet-
scented in the evening and early morning, in few-flowered racemes. Calyx
usually glandular, subequally lobed, posterior lobe smaller, often deflexed.
Corolla-tube long, cylindrical or narrowly funnel-shaped, straight or curved,
slightly widened at the throat, spurred or gibbous at the base ; spur slender,
variable in length ; limb spreading at right angles to the tube, glabrous to
subtomentose on the upper surface, subequally lobed ; lobes entire or fringed.
Stamens subequal ; filaments adnate to the corolla-tube below, free upper-
part short or very short ; anthers large, oblong to subovate, parallel, dorsifixed ; connective often apiculate ; pollen in tetrads. Ovary 2-celled, divided
by false septa into 4 compartments ; disc swollen, asymmetric, sometimes
inconspicuously so ; ovules numerous. Capsule woody, oblong to obovate,
compressed. Seeds numerous, compressed, winged, suborbicular to transversely oblong.

Corolla-lobes entire, limb pubescent above ; branchlets
and leaves glabrous ; filaments 6–8 mm. free ; seeds
(incl. wing) suborbicular, ± equally winged all round . 1. *S. rivae*
Corolla-lobes fringed, limb glabrous above ; branchlets
and leaves pubescent ; filaments almost completely
adnate ; seeds (incl. wing) transversely oblong, wing
broadest at the sides   .   .   .   .   .   . 2. *S. busseanus*

1. **S. rivae** *Engl.* in Ann. Ist. Bot. Rom. 7 : 30 (1897) ; F.T.A. 4 (2) : 569
(1906). Type : Ethiopia, Ogaden desert, near Karoul, *Riva* 1657 (FI, holo. !)

Shrub or small tree 2–6 m. high ; trunk somewhat swollen at the base ;
branches virgate, pale brown or cinereous, sparsely spiny, glabrous, rarely
shortly and sparsely pubescent on the young shoots ; spines straight or
rarely slightly recurved, 5–9 mm. long, swollen at the base. Leaves usually
shortly petiolate, obovate (occasionally ovate), 2–8 cm. long, 1·3–6 cm.
broad, usually cuneate at the base, rounded or emarginate at the apex,
glandular on both surfaces very densely so below, otherwise glabrous.
Flowers white, whitish-brown or white with a reddish spur and tube (Fig.
1/2) ; pedicels up to 1 cm. long. Calyx-lobes glandular, 2–2·5 mm. long,

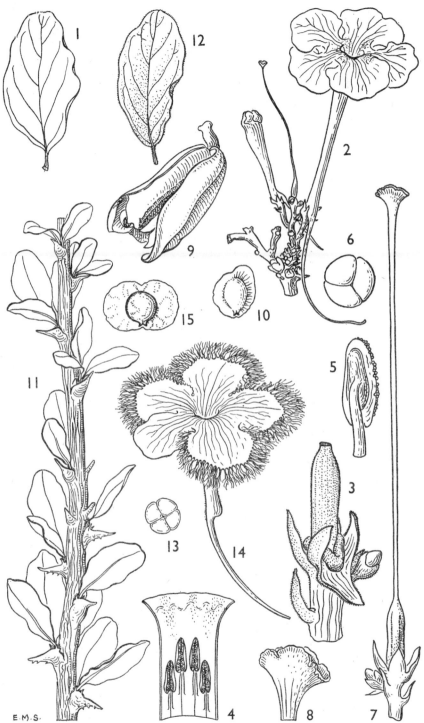

FIG. 1. *SESAMOTHAMNUS RIVAE*—1, leaf, upper surface, × 1 ; 2, inflorescence, × ⅔ ; 3, ovary, disc & calyx, × 3 ; 4, part of corolla-tube opened out to show stamens, × 1½ ; 5, anther, from back, × 4 ; 6, pollen tetrad, × 70 ; 7, calyx with pistil, × 2 ; 8, stigma, × 6 ; 9, capsule, × 1 ; 10, seed, × 1½. *S. BUSSEANUS*—11, leafy branch, × ⅔ ; 12, leaf, upper surface, × 1 ; 13, mucilage-gland, × 200 ; 14, corolla, × ⅔ ; 15, seed × 1½.—1, 9, 10 from *Dale* 3877 ; 2, from *Bally* 5656 ; 3–8, from *Gardner* 2958 ; 11–13, from *Burtt* 4999 ; 14 from *Bally* 7889 ; 15 from *Burtt* 4757.

oblong to ovate, obtuse or subacute. Corolla-tube 5–7 cm. long, narrowly cylindrical below, 2·5 mm. diameter, widened to about 6 mm. at the throat, pubescent within near the base and in the throat, otherwise glabrous ; spur 3–6 cm. long, very slender, straight or slightly curved ; limb 3–6 cm. in diameter, thinly pubescent to subtomentose on the upper surface, particularly towards the centre, glandular below ; lobes not fringed, transversely elliptic to subquadrate, undulate, 1–2 cm. long, 1·5–3 cm. broad. Filaments free for 6–8 mm., adnate below ; anther-cells subequal, not apiculate. Capsule 3·5–6·5 cm. long, varying from apiculate to emarginate at the apex, 2–2·5 cm. broad. Seeds (Fig. 1/10) obovate, ± equally winged all round, 4–5 mm. broad, 5–6 mm. long excluding the 1·5–2·5 mm. broad wing. Fig. 1/1–10, p. 3.

KENYA. Northern Frontier Province : Mt. Kulal, Elgijada, Sept. 1944, *Bally* 3452 ! & Dandu, 16 Mar. 1952, *Gillett* 12554 ! ; Teita District : Ndi, Jan. 1938, *Dale* 3877 ! TANGANYIKA. Moshi District : Kilimanjaro, May 1893, *C. S. Smith* ! ; Pare District : Kisiwani, Feb. 1936, *Greenway* 4565 !
DISTR. **K**1–3, 7 ; **T**2, 3 ; Ethiopia and Somaliland
HAB. *Acacia-Commiphora* scrub, on laval, sandy or rocky soil ; 170–1110 m.

SYN. *S. erlangeri* Engl. in E.J. 32 : 113 (1902). Type : Ethiopia, Arusi Galla country, Wabi-Budugo, *Ellenbeck* 1166 (B, holo. †)
     S. smithii [Bak. ex] Stapf in F.T.A. 4 (2) : 568 (1906) ; T.S.K. 160 (1936) ; T.T.C.L. 2 : 449 (1949). Type : Tanganyika, Kilimanjaro, *C. S. Smith* (K, holo. !)

VARIATION. Some specimens from the northern area of distribution (Somaliland, Ethiopia and northern Kenya) have shorter internodes and seeds which tend to be transversely oblong, possessing a slightly broader side wing. In *Gillett* 14048, from near Moyale, Northern Kenya, the leaves low down on the bush are pinnately lobed, whilst the upper ones are entire.

2. **S. busseanus** *Engl.* in E.J. 32 : 114 (1902) ; F.T.A. 4 (2) : 567 (1906) ; T.T.C.L. 2 : 449 (1949). Types : Tanganyika, south of Lake Victoria, Salanda, *Fischer* 454 (B, syn. †) ; Dodoma District, near Ipala in Ugogo, *Busse* 222 (B, syn. †)

Shrub or small tree 2–5 m. high with a swollen trunk and soft wood ; bark dark coppery-green scaling off in papery shavings ; branches pale brown to reddish brown, glabrous to densely pubescent particularly on the young shoots ; spines numerous, straight or slightly recurved, swollen at the base, up to 1·5 cm. long, usually bearing a pair of smaller spines near the base on the upper side. Leaves shortly petiolate, obovate, 2–5 cm. long, 1–2·5 cm. broad, densely glandular below, shortly pubescent on both surfaces or glabrescent above. Flowers (Fig. 1/14) white, or white with a crimson tube ; pedicels about 5 mm. long. Calyx-lobes glandular and pubescent, triangular, acute. Corolla-tube 2·5–4 cm. long, 3–5 mm. broad at the base widening to 7–11 mm. at the throat, glabrous to glandular-pubescent without, glabrous within ; spur slender, 4–6 cm. long ; limb 4–9 cm. in diameter, glabrous above, sparsely glandular below ; lobes subequal, all except outer one conspicuously fringed, obovate to transversely elliptic, 1·2–2 cm. long, 1·5–3 cm. broad, excluding the 1–2 cm. deep fringe. Filaments free for 1·5 mm., adnate below ; anthers apiculate. Capsule 3·5–4 cm. long, 1·6–2 cm. broad. Seeds (Fig. 1/15), including the wing, transversely oblong ; wing 3·5–5 mm. broad at the sides, narrow at the base and apex. Fig. 1/11–15, p. 3.

KENYA. Northern Frontier Province : Mandera, 24 May 1952, *Gillett* 13307 ! ; Tana River District : Garissa road, Golana Gof, May 1945, *Bally* 4375 !
TANGANYIKA. Dodoma District : Manyoni, Dec. 1933, *B. D. Burtt* 4999 ! ; Mbeya District : between Ikoga & Isunura, Sept. 1936, *B. D. Burtt* 6008 !
DISTR. **K**1, 4, 7 ; **T**5, 7 ; Somaliland
HAB. *Acacia-Commiphora* scrub, rather more xerophytic than *S. rivae* but the two species overlap considerably ; 300–1350 m.

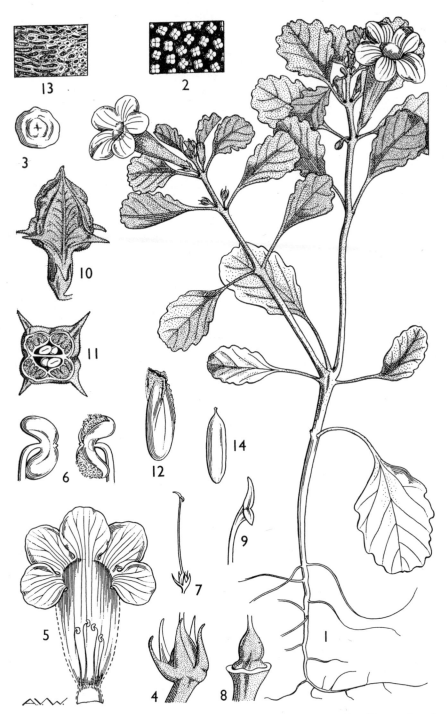

FIG. 2. *PEDALIUM MUREX*, mainly from *Napier* 6296—1, plant, × 1 ; 2, portion of lower surface of
leaf with mucilage-glands, × 40 ; 3, nectarial gland from base of pedicel, × 24 ; 4, calyx, × 6 ; 5, corolla,
opened out, × 2 ; 6, anthers, × 16 ; 7, pistil and calyx, × 2 ; 8, ovary and disc, × 6 ; 9, stigma, × 6 ;
10, fruit, × 2 ; 11, fruit, t.s., × 2 ; 12, seed, × 4 ; 13, portion of testa, × 20 ; 14, seed, with testa
removed, × 4.

Variation.   Specimens from the northern area of distribution (Kenya & Somaliland) have leaves and branchlets which are only sparsely pubescent and corolla-tubes which are glabrous except for the mucilage-glands, though in one case (*Gillett* 12574) one or two hairs are visible, whereas specimens from the south (Tanganyika) have densely pubescent leaves and branchlets and corolla-tubes which are pubescent with multi-cellular hairs in addition to the mucilage-glands, though in one case (*B. D. Burtt* 6008) the hairs are not very conspicuous.   At first it appeared that the species might be separated into two varieties but on further examination it was decided that the differences were more of degree and that it would be better to treat the whole as one variable species.

2. PEDALIUM
L., Syst. Nat., ed. 10, 1123 (1759)

Annual herb with simple or branched subsucculent stem.   Leaves petiolate, entire or coarsely dentate.   Flowers solitary in the leaf axils, with nectarial glands at the base of the pedicels.   Calyx small.   Corolla-tube subcylindric or narrowly funnel-shaped; limb spreading subequally 5-lobed. Stamens included; staminode often present; anthers large, divergent, kidney-shaped.   Ovary 2-celled, cells undivided; 2 ovules in each cell; disc small, inconspicuous.   Fruit indehiscent, subpyramidal, 4-angled, rounded to acute at the apex with a spreading spine at each basal angle, and then abruptly contracted into a narrow collar below.   Seeds 1–2 in each cell.

P. murex *L.*, Syst. Nat., ed. 10, 1123 (1759) ; F.T.A. 4 (2) : 540 (1906) ; W.F.K. 99 (1948).   Type : India, no collector (LINN, lecto. !)

Erect, spreading or subprostrate, sparsely glandular herb, 12–75 cm. high. Leaves elliptic, obovate or oblong sometimes fleshy ; petiole 0·5–3·5 cm. long, blade 1·5–5 cm. long, 0·8–3·5 cm. broad, coarsely dentate, particularly in the upper half, rounded or truncate at the apex.   Flowers pale primrose-yellow.   Calyx-lobes lanceolate, acuminate.   Corolla-tube 2–2·5 cm. long, glandular ; limb 1·5–2 cm. in diameter, glabrescent or with a few hairs in the throat ; lobes suborbicular.   Fruit 1–2 cm. long, 0·6–1 cm. broad, excluding the spines, sparsely glandular, rugose or tuberculate on the face.   Seeds (Fig. 2/12) elongate-elliptic, 6 mm. long, 1·5 mm. broad, 3-angled towards the apex.   Fig. 2, p. 5.

Kenya.   Mombasa District : English Point, 26 May 1934, *Napier* 3265 !
Tanganyika.   Rufiji District : Mafia Island, 3 Apr. 1933, *Wallace* 746 !
Zanzibar.   Near Zanzibar, 9 Feb. 1929, *Greenway* 1385 !
Distr.   K1, 7 ; T3, 6, 8 ; Z ;   Gold Coast, Somaliland, Ethiopia, Socotra, Portuguese East Africa & Madagascar ; also in India and Ceylon
Hab.   A saline soil indicator in sand or on limestone in short grass near the coast, also in old sisal plantations ;   0–440 m.

Variation.   A specimen named *P. macrocarpum* Engl. (*nomen nudum*) from Mombo, Usambara Mts. (*Zimmermann* 953) probably belongs to the above species though it has longer, narrower leaves and very short spines to the fruit.

Note.   The leaves are eaten as a vegetable.

3. PTERODISCUS
Hook., Bot. Mag. t. 4117 (1844)
Pedaliophyton Engl. in E.J. 32 : 111, t. 5 (1902)

Small perennial herbs or occasionally small shrubs rarely more than 30 cm. high, often subsucculent with a swollen stem-base and tuberous root from which arise one to many suberect simple or branched stems.   Leaves variable in shape, entire, undulate, dentate or pinnatifid.   Flowers solitary in the leaf-axils, variable in colour.   Calyx small.   Corolla-tube funnel-shaped, sometimes slightly gibbous at the base ; limb spreading sub-bilabiate ; lobes subequal, ovate, orbicular or transversely elliptic.   Stamens included,

anther-cells large, divergent, pendulous, dehiscing by a short slit.   Ovary
2-celled, cells undivided ; ovules 1–3 or 10–12 in each cell; disc unilateral
(sometimes inconspicuously so).   Fruit indehiscent, 4-winged or 4-angled,
often ± compressed.   Seeds variable in shape.

Fruits ± subconical, 4-angled or very narrowly 4-winged,
    scarcely compressed ;   seeds elongate ellipsoid ;
    ovules 1-seriate, 1–3 in each cell :
    Flowers blue or mauve and white, corolla-tube
       1–1·4 cm. long ; leaves oblong, 1·3–4 cm. long,
       0·4–1·5 cm. broad, rounded or subtruncate at
       the apex ; ovules 2–3 in each cell .   .   .   1. *Pt. coeruleus*
    Flowers yellow or orange, corolla-tube 2–4 cm.
       long ; leaves narrowly oblong-lanceolate, 2·5–
       13 cm. long, 0·6–1·2 cm. broad, subacute at the
       apex ; ovules 1–2 in each cell   .   .   .   2. *Pt. angustifolius*
Fruits suborbicular, conspicuously 4-winged, com-
    pressed ;   seeds flattened, suborbicular ;   flowers
    yellow, corolla-tube 1·6–3 cm. long ;   leaves
    obovate to ovate, 1·5–5·5 cm. long, 0·8–3 cm.
    broad ; ovules 2-seriate, 5–6 in each series.   .   3. *Pt. ruspolii*

1. **Pt. coeruleus** *Chiov.*, Fl. Somala 2 : 342 (1932).   Type : Somaliland,
Upper Juba, Obe, *Senni* 545 (FI, holo. !)

Plant 5–20 cm. high, simple or branched.   Leaves 0·4–2·5 cm. petiolate ;
blade oblong 1·3–4 cm. long, 0·4–1·5 cm. broad, rounded or subtruncate at
the apex, cuneate at the base, margins undulate, thinly glandular below,
glabrous above.   Flowers blue, or white suffused mauve in the throat with
mauve veins and vermilion edges to the lobes.   Corolla-tube 1–1·4 cm. long,
glabrous within ;   limb about 1 cm. diameter, lobes suborbicular about
4 mm. long and broad, not ciliate.   Filaments about 4 mm. long with a few
hairs at the base.   Ovary ovoid, puberulous ;   disc small, unilateral ; ovules
2–3 in each cell.   Fruit subconical or heart-shaped, 4-angled or very narrowly 4-winged, 1·2–1·5 cm. long, 1–1·2 cm. broad at the base, acute or subacute at the apex, subauriculate at the base.   Seeds elongate-ellipsoid,
about 5 mm. long, 2 mm. broad.

Kenya.   Northern Frontier Province : Bura-Ijara road, mile 9, 7 Jan. 1943, *Bally*
   2058 !
Distr.   **K**1 ; Somaliland
Hab.   Semi-desert scrub ; 100 m.

2. **Pt. angustifolius** *Engl.* in E.J. 19 : 155 (1894) ; E. & P. Pf. 4 (3b) :
260 (1895); F.T.A. 4 (2) : 545 (1906) ; E. A. Bruce in K.B. 1953 : 419
(1953).   Type : Tanganyika, Mwanza District, between Mago and Kagehi,
Fischer 462 (B, holo.†, K, iso. !—1 flower & leaf)

Plant branched just above the base ; branches spreading, more or less
fleshy, 9–20 cm. long, purplish, glabrous, fairly densely leafy.   Leaves more
or less fleshy, petiolate, ascending ; petiole 0·5–2·5 cm. long ; blade dark
green, narrowly oblong-lanceolate, lanceolate or narrowly oblanceolate,
2·5–13 cm. long, 0·6–1·2 cm. broad, sparsely glandular when young otherwise glabrous, very gradually cuneate at the base, obtuse or subacute at the
apex, margins entire or undulate, very rarely toothed towards the apex,
often wrinkled below (in the dried state).   Flowers yellow or orange, sometimes with a purple blotch in the tube, often sweet-scented, pedicels 1·5–3
mm. long, glandular.   Corolla-tube straight or slightly curved, 2–4 cm.
long, sparsely glandular without, slightly oblique at the mouth, pubescent
with multicellular hairs within in the upper part particularly in the throat ;

lobes ciliate, varying in shape from ovate to transversely elliptic, upper one larger, 0·7–1·2 cm. long, 0·6–1·5 cm. broad, laterals 0·6–1 cm. long, 5–8 mm. broad, lower 5–7 mm. long, 8–10 mm. broad (badly pressed flowers appear smaller), pubescent within, glabrous outside. Filaments 4–9 mm. long, all hairy at base, longer 2 pubescent to the top. Ovary ovoid, disc small inconspicuously unilateral, ovules 1–2 in each cell. Fruit ovoid to conical, slightly compressed, 4-angled, obtuse to subacute at the apex, angles narrowly winged, sometimes inconspicuously so, faces tuberculate, 1–1·5 cm. long, 0·8–1 cm. broad at the base (Fig. 3/19). Seeds (Fig. 3/21) elongate-elliptic, about 6·5 mm. long, 2 mm. broad, with a narrow collar at the apex. Fig. 3/14–21.

TANGANYIKA.   Shinyanga District : Seseku and Mhumbo Valley, Jan. 1937, *B. D. Burtt* 6430 !
DISTR.   T1, 8 ; endemic in Tanganyika
HAB.   In open ground, on sandy, grey clay or seasonally water-logged soil, often in *Commiphora-Acacia* bush ; 1200–1300 m.

SYN.   *Pedaliophyton busseanum* Engl. in E. J. 32 : 111, t. 5 (1902). Type : Tanganyika, near Ruvuma River, Kwa-Mbaramula, *Busse* 1046 (B, holo. †, EA, iso. !)
   Pedalium busseanum (Engl.) Stapf in F.T.A. 4 (2) : 540 (1906)

NOTE.   This species appears to be quite abundant over a small area south of Lake Victoria ; from here there is a big jump in the distribution to the Ruvuma River on the borders of Tanganyika and Portuguese East Africa. It is strange that no material has been received from intervening localities.

3. **Pt. ruspolii** *Engl.* in Ann. Ist. Bot. Roma 7 : 31 (1897) ; F.T.A. 4 (2) : 544 (1906) ; E. A. Bruce in K.B. 1953 : 420 (1953). Types : Italian Somaliland, R. Daua near Dolo, *Ruspoli & Riva* 1101 (FI, syn. !, K, photo !), & R. Daua near Hamae [or Hamore or Hamare], *Ruspoli & Riva* 1089 * (FI, syn. !, K, photo !)

Plant with a fleshy base 4–8 cm. long, 0·5–2 cm. diameter from which up to 20 stems arise, which are erect or suberect, 4–20 cm. long. Leaves variable in shape, usually long-petiolate (0·5–3·5 cm.) ; blade obovate, elliptic or ovate, 1·5–6·5 cm. long, 0·8–3·5 cm. broad, rounded at the apex, cuneate at the base, thinly to densely glandular below, glabrescent above, margins entire or undulate. Flowers bright yellow to orange, sometimes with a red or purple centre spot, opening 2–3 at a time ; pedicels 2–5 mm. long, glandular. Corolla-tube straight or slightly curved, oblique at the throat, 1·6–3 cm. long, 3–4 mm. diameter, glandular without, pubescent with multicellular hairs within in the upper part particularly in the throat ; limb 1·5–3·5 cm. diameter, pubescent near the centre ; lobes more or less transversely elliptic to suborbicular, 0·6–1 cm. long, 1–1·5 cm. broad, often ciliate. Filaments bearded at the base, the 2 longer pubescent throughout, about 7 mm. long. Ovary ovoid or oblong-ovoid ; disc conspicuous ; ovules compressed, overlapping, 2-seriate, about 5–6 in each row. Fruit suborbicular, 4-winged, 1·7–3 cm. diameter, central portion more or less inflated, leathery, not woody (Fig. 3/11). Seeds (Fig. 3/13) compressed, obovate to suborbicular, about 5 mm. diameter, central portion rather inflated, foveolate or subrugose, margins thin. Fig. 3/1–13.

KENYA.   Lake Rudolf, on central island, 1 May 1934, *Martin* 141 ! ; Meru District : Isiolo, 21 Dec. 1943, *Bally* 3709 !
DISTR.   K1, 2, 4 ; Somaliland, Ethiopia & A.-E. Sudan
HAB.   Common in open semi-desert scrub and grassland, on dry sandy or rocky ground and alluvial soil ; 150–1160 m.

SYN.   *P. somaliensis* [Bak. ex] Stapf in F.T.A. 4 (2) : 544 (1906) ; W.F.K. 99 (1948). Type : Somaliland Protectorate, Harradigit, *James & Thrupp* (K, holo. !)
   P. wellbyi Stapf in F.T.A. l.c. Type : Kenya, Turkana District, Turkwell River, SW. of Lake Rudolf, *Wellby* (K, holo. !)

* See E. A. Bruce in K.B. 1953 : 419, footnote 4 (1953)

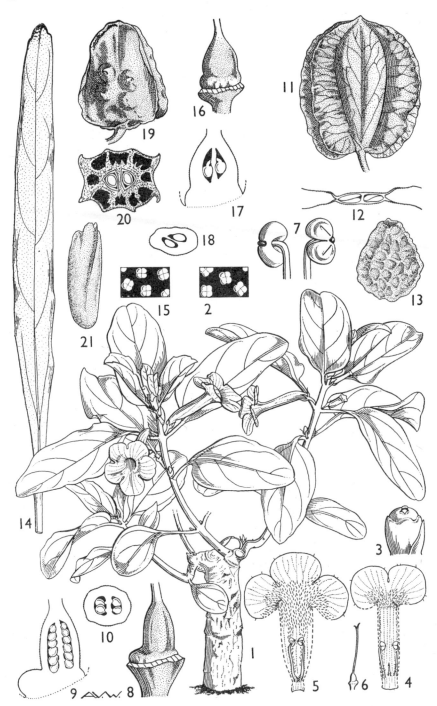

FIG. 3. *PTERODISCUS RUSPOLII*, mainly from *Gillett* 15130—1, plant, × ½ ; 2, portion of lower surface of leaf with mucilage-glands, × 40 ; 3, nectarial gland, × 16 ; 4 and 5, corolla, l.s., × 1 ; 6, calyx and pistil, × 1 ; 7, anthers, × 6 ; 8, ovary and disc, × 6 ; 9 and 10, ovary and ovules, l.s. and t.s., × 8 ; 11, fruit, × 2 ; 12, fruit, t.s., × 2 ; 13, seed, × 4. *PT. ANGUSTIFOLIUS*, from *Burtt* 3520 and 6430— 14, leaf, × 1 ; 15, portion of lower surface of leaf, × 40 ; 16, ovary and disc, × 6 ; 17 and 18, ovary and ovules, l.s. and t.s., × 8 ; 19, fruit, × 2 ; 20, fruit, t.s., × 2 ; 21, seed, × 4.

4. DICEROCARYUM

Boj. in Ann. Sci. Nat., sér. 2, 4 : 268 (1835) ; Merrill in Trans. Am. Phil. Soc.,
n.s. 24, 2 : 355 (1935) ; E. A. Bruce in K.B. 1953 : 421 (1953)

Pretrea [J. Gay ex] Meisn., Pl. Vasc. Gen. 1 : 298 (1836–43)

Perennial herb with trailing procumbent branches, very variable in leaf-shape, pubescence and indentation. Flowers solitary in the leaf-axils, with nectarial glands at the base of the pedicels. Corolla-tube obliquely campanulate, curved, limb sub-bilabiate, 5-lobed, upper and lateral lobes sub-equal, lowest lobe longest. Anthers oblong, dorsifixed ; filaments filiform, free from the corolla-tube. Ovary slightly compressed, 2-celled and 2-horned at the apex, cells completely divided by a false septum into 4 compartments each containing 2 ovules. Fruit woody, indehiscent, more or less disc-shaped with a raised 2-horned central portion on the upper surface. Seeds 2 in each compartment.

D. zanguebarium (*Lour.*) *Merrill*, 1 c. Type : Portuguese East Africa, near Mozambique Is., *Loureiro* (P, holo. !, K, photo !)

Procumbent herb with trailing stems 0·45–2·1 m. long, sparsely to densely pubescent. Leaves oblong, ovate or cordate-ovate in outline, grossly dentate to pinnately incised, 1·1–5 cm. long, 0·6–4 cm. broad, usually discolorous, thinly to densely pubescent on both surfaces, glandular below; petiole 0·3–1·6 cm. long (Fig. 4/1–7). Flowers pink, mauve, violet or crimson often streaked or dotted with a darker colour ; pedicels 1·5–5·5 cm. long. Calyx-lobes lanceolate, pubescent. Corolla pubescent, varying from 1·5–4·5 cm. in length ; lobes rounded at the apex (Fig. 4/10). Ovary densely pubescent ; disc saucer-shaped. Fruit 1·5–2·6 cm. long, 1–2 cm. broad, pubescent; horns about 5 mm. long (Fig. 4/14–15). Seeds (Fig. 4/16) subovoid, reticulate, slightly 4-angled, scarcely compressed, 4 mm. long, 2·5 mm. broad at the base. Fig. 4.

TANGANYIKA. Pangani District : Bushiri Estate, July 1950, *Faulkner* 584 ! ; Uzaramo District : Dar es Salaam, Feb. 1874, *Hildebrandt* 1133 ! ; Rufiji District : Mafia Island, Kilindoni, Sept. 1937, *Greenway* 5239 !
ZANZIBAR. Unlocalized, *Bojer*! ; Zanzibar Is., Mnazi Mmoja, Aug. 1951, *R. O. Williams* 80 !
DISTR. **T**3, 6 ; **Z** ; Portuguese East Africa, Northern and Southern Rhodesia, Angola, Bechuanaland, Transvaal & South West Africa

HAB. A typical strand plant on sandy ground by sea shore, sometimes in association with *Dactyloctenium* and *Ipomoea biloba* ; 0–500 m.

SYN. *Martynia zanguebaria* Lour., Fl. Coch. 386 (1790). Type : Portuguese East Africa, near Mozambique Is., *Loureiro* (P, holo. !, K, photo. !)
Dicerocaryum sinuatum Boj. in Ann. Sci. Nat., sér. 2, 4 : 269, t. 10 (1835). Type : Zanzibar, *Bojer* (P, holo., K, iso. !)
Pretrea zanguebaria (Lour.) [J. Gay ex] DC., Prod. 9 : 256 (1845) ; F.T.A. 4 (2): 565 (1906) [as *zanguebarica*]

VARIATION. There is considerable variation within this species, particularly in leaf-shape, indentation, indumentum, length of pedicel and flower-size. The species described within this " group " (now reduced to synonymy) were differentiated from one another by characters which are not constant, e.g. some specimens from the interior (Bechuanaland, Amboland and Ngamiland) described under *Pretrea eriocarpa* Decne., appear to differ from the coastal forms in having shorter, broader, less deeply divided leaves ; there are, however, intermediate forms, particularly from the Transvaal and Rhodesia, which have both leaf-forms. Other characters have been tried out but appear to vary independently, so that the species cannot be separated into clear-cut varieties. In view of this lack of correlation it has been decided to treat the whole as one variable species.

NOTE. The citation " British East Africa : near Malindi, *Allen* 157," in F.T.A. 4 (2): 566 is an error. C. E. F. Allen collected in Southern Rhodesia and his number·157 was collected there near Malindi (18° 45′ S., 27° 1′ E.) in November 1905.

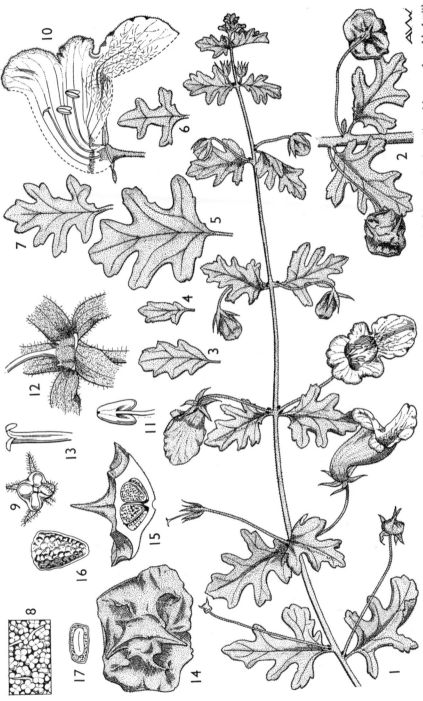

FIG. 4. *DICEROCARYUM ZANGUEBARIUM*—1, flowering branch, × 1; 2, part of fruiting branch, × 1; 3-7, leaves, × 1; 8, portion of lower surface of leaf with hairs and mucilage-glands, × 40; 9, nectarial gland at base of pedicel, × 8; 10, flower, l.s., × 2; 11, anther, from back, × 4; 12, ovary, disc, part of style and calyx, × 6; 13, stigma, × 6; 14, fruit, from above, × 2; 15, fruit, l.s., × 2; 16, seed, × 4; 17, seed, t.s., showing endosperm and embryo, × 4.—1, 2, 8-13, from *Faukner* 584; 3, 16, 17, from *Greenway* 5239; 4, 6 from *Greenway* 1330; 5, from *Schlieben* 2576; 7, 14, 15, from *Burtt* 5188.

5. JOSEPHINIA

Vent., Jard. Malm. 67, t. 67 (1804) ; E. A. Bruce in K.B. 1953 : 422 (1953)

Pretreothamnus Engl. in E.J. 36 : 228 & Fig. (1905)

Erect or decumbent herbs or small shrubs. Leaves opposite or sub-opposite, petiolate, lanceolate, elliptic or obovate, entire, dentate or lobulate. Flowers solitary in the upper leaf-axils. Calyx-lobes persistent, subequal or the posterior smaller. Corolla-tube shortly tubular at the very base then inflated and obliquely campanulate ; limb sub-bilabiate ; upper lip suberect, shortly 2-lobed ; lower lip subhorizontal, 3-lobed with the central lobe the longest. Anther-cells parallel ; connective gland-tipped. Ovary sub-verrucose, 2-4 celled ; cells divided by false septa into 4-8 compartments with 1-3 erect ovules in each compartment ; disc circular, rather unequally thickened. Fruit indehiscent, ovoid to subglobose, beaked or not, hard and woody, densely covered with stout spines, divided into 4-8 compartments each containing 1-3 seeds. Seeds erect, oblong to elongate-ovoid, sometimes subtriquetrous at the base ; testa smooth or wrinkled.

Josephinia africana *Vatke* in Linnaea 43 : 541 (1882) ; Chiov., Fl. Somala 2 : 344 (1932). Type : Kenya, Teita District, Tsavo River, *Hildebrandt* 2586 (B, holo. †)

Low shrub or semi-decumbent woody herb 30–130 cm. high ; branches opposite, dark brown, densely crisped-pubescent to pilose when young, becoming glabrous. Leaves pale green ; petiole 2–10 mm. long ; blade obovate, lobulate or coarsely sinuate-dentate, 1·3–2·5 cm. long, 0·5–1 cm. broad, rounded at the apex, cuneate at the base, pubescent and glandular on both sides, more densely so below, becoming glabrous. Pedicels 4–7 mm. long, densely pubescent. Calyx-lobes oblong, subequal, about 5 mm. long, 1 mm. broad, rounded or subacute at the apex, pubescent and ciliate. Cor-olla mauve to pale pink, spotted with black inside, thinly pubescent and glandular, 2·6–4 cm. long, about 2–3·5 cm. diameter across the limb ; central lobe of lower lip suborbicular to transversely elliptic, 1·5–2 cm. across, others much shorter, all rounded at the apex. Stamens with a tuft of hairs at the base of the filaments. Ovary ovoid to subglobose, densely pubescent or pilose, about 2 mm. long, 1·5 mm. diameter, 2-celled, divided into 4 com-partments by a false septum, with 3 superposed small ovules in each com-partment ; style flattened about 1·3 cm. long ; disc fleshy. Fruit sub-globose, pubescent, not beaked, 1·5–2 cm. diameter including the 3–4 mm. long spines, divided into 4 compartments each containing 2–3 seeds (Fig. 5/10 & 11). Seeds elongate-ovoid, slightly triquetrous towards the base ; testa wrinkled or reticulate. Fig. 5.

KENYA. Northern Frontier Province : S. of El Wak, 26 May 1952, *Gillett* 13346 ! ;
    Teita District : Voi, 10 May 1931, *Napier* 1028 !
DISTR. **K**1, 7.; Southern Somalia
HAB. In dry *Acacia-Commiphora* and semi-desert scrub and on the margins of sisal
    plantations ; 100–650 m.

SYN. *Pretreothamnus rosaceus* Engl. in E.J. 36 : 228 (1905). Type : Kenya, Northern
        Frontier Province, Jeroko [SW. of Mandera], *Ellenbeck* 2199 (B, holo. †)
    P. africanus (Vatke) B. Fedtsch. in Bull. Jard. Imp. Bot. Pétersb. 15 : 404 (1915)

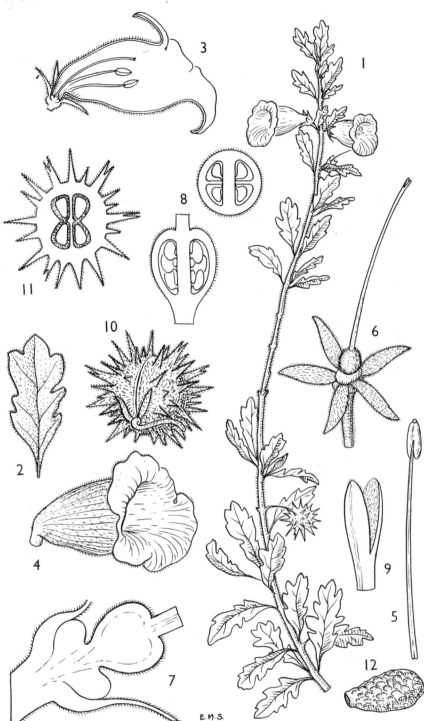

E.M.S.

FIG. 5. *JOSEPHINIA AFRICANA*—1, branch with flowers and fruits, × ⅔ ; 2, leaf, upper surface, × 2 ; 3, flower, l.s., × 1½ ; 4, corolla, × 1½ ; 5, stamen, × 4 ; 6, calyx with disc, and pistil, × 4 ; 7, diagram of ovary and disc, l.s., × 16 ; 8, diagrams of ovary, t.s. and l.s. ; 9, stigma, × 24 ; 10, fruit, × 3 ; 11, fruit, t.s., × 3 ; 12, seed, × 6.—1, 2, 10, 11, from *Napier* 1028 ; 3–9, 12, from *Gillett* 15150.

6. CERATOTHECA

Endl. in Linnaea 7 : 5, tt. 1, 2 (1832)

Erect or procumbent, annual or perennial herbs. Leaves membranous, petiolate, often polymorphic. Flowers solitary in the leaf-axils, shortly pedicellate, red, pink, lilac, mauve or yellow, with nectarial glands at the base of the pedicels (Fig. 6/6 & 7). Calyx often persistent. Corolla obliquely campanulate; limb sub-bilabiate; upper and lateral lobes short subequal, lowest lobe the longest. Filaments slender; anther-cells oblong, parallel. Disc annular, subequal. Ovary subcylindrical, slightly compressed, 2-celled; cells divided by a false septum almost to the apex; ovules numerous, 1-seriate in each compartment. Capsule oblong, compressed, rounded or truncate at the apex and horned at the angles (horns rarely suppressed), splitting loculicidally. Seeds numerous, compressed, obovate to suborbicular with a narrow margin and a smooth or wrinkled centre portion.

Leaves truncate, widely cuneate or subhastate at the base,
    longer than broad, about 1·5–8 cm. long, 0·4–4·5 cm.
    broad ; petiole usually shorter than the blade ;
    bracteoles linear, shorter than the pedicel   .   .   1. *C. sesamoides*
Leaves narrowly cordate at the base, about as broad as
    long, 3·5–8 cm. long, 3–8 cm. broad ; petiole usually
    longer than the blade ; bracteoles linear-filiform,
    conspicuous, longer than the pedicel   .   .   .   2. *C. sp.*

1. **C. sesamoides** *Endl.* in Linnaea 7 : 5, tt. 1, 2 (1832) ; F.T.A. 4 (2) : 563 (1906) ; F.W.T.A. 2 : 244 (1931). Type : Senegal, *Kohaut* (B, holo. †, K, iso. !)

Simple or branched, erect or semi-prostrate, pubescent, annual herb, 15–90 cm. high. Leaves usually opposite, varying in shape from ovate-deltoid to lanceolate-deltoid and more rarely ovate-cordate, upper ones narrower sometimes subentire, lower ones usually coarsely dentate at least towards the base (Fig. 6/2–4), 1·5–8 cm. long, 0·4–4·5 cm. broad, acute at the apex, truncate, widely cuneate or subhastate at the base, pubescent, densely glandular below, very sparsely so above ; petioles pubescent, variable in length, lower ones up to 6 cm., upper ones from 0·1 cm. Flowers pink, lilac, mauve or purple, throat and lip often cream with darker lines ; pedicels 3–5 mm. long. Calyx persistent, lobes lanceolate, pubescent, acute, 4–5 mm. long. Corolla 1·3–4 cm. long, thinly pubescent ; lowest lobe broadly ovate, upper and lateral lobes much shorter, transversely elliptic. Ovary thinly pubescent. Capsule pubescent and glandular, 1·2–2·3 cm. long, 0·4–0·7 cm. broad ; horns varying from 1–3·5 mm. long, occasionally suppressed (Fig. 6/12–14). Seeds (Fig. 6/15 & 16) pale brown or black, 2·5–4 mm. long, 2–2·5 mm. broad, margin transversely rugose, central portion convex, smooth. Fig. 6.

UGANDA. Acholi District : near Rom, Oct. 1947, *Dale* U484 !
KENYA. Masai District : Mpagazi River, Nguruman Escarpment, 25 Sept. 1944, *Bally* 3813 !
TANGANYIKA. Shinyanga, May 1935, *B. D. Burtt* 5121 ! ; Ufipa District : Tumba River, 13 March 1951, *Bullock* 3771 !
DISTR. U1 ; K6 ; T1, 3–6 ; West Africa, A.-E. Sudan, Portuguese East Africa, Nyasaland, Northern and Southern Rhodesia
HAB. A weed of native cultivations and wasteland near villages, on sandy soil ; 800–1600 m.

SYN. *Ceratotheca melanosperma* [Hochst. ex] Bernh. in Linnaea 16 : 32 & 41 (1842). Type : A.-E. Sudan, Kordofan, Arasch-Cool Mountains, *Kotschy* 101 (K, iso. ! BM, iso. !)
    C. sesamoides Endl. var. *melanoptera* A.DC. in DC., Prodr. 9 : 252 (1845). Type as in *C. melanosperma*.

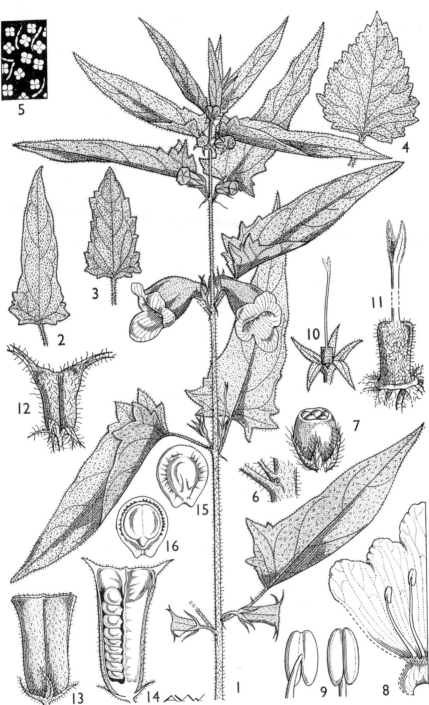

FIG. 6. *CERATOTHECA SESAMOIDES*—1, part of flowering and fruiting stem, × 1 ; 2–4, leaves, × 1 ;
5, portion of lower surface of leaf showing mucilage-glands and hairs, × 40 ; 6, base of pedicels and
petiole showing nectarial gland, × 2 ; 7, nectarial gland, × 16 ; 8, portion of corolla, opened out, × 2 ;
9, anther, × 6 ; 10, pistil and calyx, × 2 ; 11, pistil and disc, × 6 ; 12, 13, fruit forms, × 2 ; 14, fruit,
l.s., showing seeds and false septum, × 2 ; 15, seed, × 6 ; 16, seed, l.s., showing embryo and endosperm,
× 6.—1, 5, 8–12, from *Pielou* 92 ; 2, from *Rounce* 285 ; 3, 6, 7, from *Tanner* 398 ; 14–16, from *Burtt* 5121.

VARIATION. There is great variation within the species in leaf-shape, dentition, density of pubescence and in the length of the horns on the capsule. *Dale* U484 and *Dawe* 860 are extreme cases with very densely pubescent, broadly ovate leaves, stouter, more woody and pubescent capsules (in which the horns are often suppressed) and larger more narrowly ovate seeds.

2. C. sp.

Herb with subquadrate, 4-grooved, pilose stem. Leaves opposite, long-petiolate ; petioles 2·5–10 cm. long, pilose ; lamina ovate-cordate, 3·5–8 cm. long, 3–8 cm. broad at the base, apex subacute, base narrowly cordate, margins coarsely dentate or crenate-dentate, thinly pilose below particularly on the nerves, becoming glabrescent above. Flowers solitary in the upper leaf-axils, subsessile or shortly pedicellate ; bracteoles conspicuous, linear-filiform, about 1 cm. long, pilose. Calyx-lobes linear-lanceolate, about 1 cm. long, pilose. Corolla-tube sub-campanulate from a curved tubular base, 2·5–4 cm. long, pubescent. Ovary compressed, oblong, truncate and slightly angled at the apex, pilose.

KENYA. Nairobi, on a building site not far from the Arboretum, Jan. 1945, *Adamson in Bally* 4598 !

NOTE. Mr. Bally of the Coryndon Museum, Nairobi, states that this plant was first collected in January 1945 on a building site and about 2 years later was again gathered from the same place but has not been seen since. The locality has now been built over.
    The above specimen unfortunately has no fruits, but the oblong, compressed ovary, which is truncate at the apex, is typical of the genus *Ceratotheca*. The plant closely resembles *C. triloba* [E. Mey. ex] Bernh. a species from southern Africa and may possibly be this. It differs from it in the longer-petiolate, ovate-cordate, dentate upper leaves.

7. SESAMUM
L., Gen. Pl., ed. 5, 282 (1754)

Annual or perennial, erect or procumbent herbs. Leaves sessile or petio-late, entire, lobed or partite, often varying on the same plant. Flowers solitary in the leaf axils, shortly pedicellate, white, pink or purple, some-times spotted within. Calyx persistent or deciduous. Pedicels usually short, with nectarial glands at the base. Corolla foxglove-like, obliquely cam-panulate, limb sub-bilabiate, lowest lobe the longest. Filaments slender ; anthers dorsifixed, cells parallel, connective gland-tipped. Disc annular, regular. Ovary 2-celled, subcylindrical ; cells divided almost to the apex by a false septum ; ovules numerous, 1-seriate in each division. Capsule oblong or sub-obconical, grooved, beaked at the apex. Seeds numerous, obovate, compressed, winged or wingless, smooth or rugose.

Seeds winged ; capsule sub-obconical, narrowed to the
    base, calyx deciduous in fruit ; stems glabrous
    except for mucilage glands ; some leaves palmately
    3–5-foliolate or -partite with narrow lobes or
    segments . . . .   1. *S. alatum*
Seeds not winged ; capsule parallel-sided, rounded at
    the base, calyx usually persistent in fruit ; stems
    usually pubescent ; leaves rarely lobed or partite,
    if so segments broad, ovate :
    Seeds smooth ; calyx persistent ; stems thinly
        pubescent to glabrescent ; lower leaves long
        petiolate, often toothed, sometimes lobed or
        partite . . . .   2. *S. indicum*

Seeds rugose or pitted ; stems pubescent ; leaves
 sessile or petiolate, never partite :
 Leaves sessile or subsessile, linear to oblong,
  usually entire ; calyx persistent :
  Leaves densely glandular below, thinly pubescent on the nerves, linear to linear-lanceolate, acute or subacute at the apex, lower
   leaves sometimes toothed ; corolla 2–
   3·5 cm. long ; capsule narrowly oblong,
   3–4 mm. broad ; seeds about 1·5 mm. long 3. *S. angustifolium*
  Leaves white tomentose below obscuring the
   glands, narrowly oblong to oblong-
   oblanceolate, usually truncate or rounded
   at the apex, never toothed ; corolla large,
   3·5–7 cm. long ; capsule oblong, 5–6 mm.
   broad ; seeds 2 mm. long . . . 4. *S. angolense*
 Leaves conspicuously petiolate, ovate-lanceolate,
  ovate-cordate or lower ones 3-lobed, usually
  cordate at the base, margins serrate ; calyx
  ultimately deciduous . . . 5. *S. latifolium*

1. **S. alatum** *Thonn.* in Schumach. & Thonn., Beskr. Guin. Pl. 284 (1827) ;
F.T.A. 4 (2) : 559 (1906) ; F.W.T.A. 2 : 243 (1931). Type : Gold Coast,
Thonning (C, holo. !)

Erect herb 20–90 (rarely 150) cm. high, stem simple or sparsely branched,
sulcate, glabrous except for the mucilage glands. Leaves generally heteromorphic, opposite or rarely alternate, lower leaves long-petiolate (2–7 cm.),
palmately 3–5-foliolate or partite ; lobes lanceolate to narrowly linear-lanceolate, 1·5–7 cm. long, 0·2–1 cm. broad, central lobe longest, cuneate at the
base, acute or rounded at the apex (Fig. 7/1) ; margins entire or undulate ;
glabrous except for the mucilage glands which are denser below ; upper
leaves usually simple, linear-lanceolate to finely linear, 3–6 cm. long, attenuate into a 1–2 cm. long petiole. Flowers pink or purple, sometimes red-
spotted within. Calyx deciduous ; lobes linear-lanceolate acuminate, 3–3·5
mm. long, densely glandular and thinly pubescent. Corolla 2–3 cm. long,
thinly glandular and pilose outside with white multicellular hairs, thinly
pubescent within. Ovary densely ascending adpressed-pubescent, slightly
compressed and 4-angled, about 5 mm. long and just over 1 mm. broad,
attenuate at the apex into a slender style. Capsule narrowly obconical,
2–4 cm. long, excluding the 4–12 mm. long beak, 5–7 mm. broad in the
upper part, gradually narrowed to the base, abruptly acuminate and beaked
at the apex, glandular and thinly pubescent, becoming glabrous, slightly
compressed and 4-sulcate (Fig. 7/2). Seeds (Fig. 7/3) obliquely overlapping
in the capsule, 2–3 mm. long, foveolate, with a suborbicular 2–3 mm. long
wing at the base and apex. Fig. 7/1–3, p. 18.

KENYA. Northern Frontier Province : Lake Rudolf, *Jex-Blake in C.M.* 6862 ! ;
 Turkwell Valley, near Kaputi, *Champion* T62 !
DISTR. K1, 2 ; Senegal, French Sudan, Gold Coast, Nigeria, A.-E. Sudan, Ethiopia,
 Southern Rhodesia, Portuguese East Africa, Bechuanaland
HAB. Semi-desert grassland on river banks ; 600–900 m.

SYN. *Volkameria alata* (Thonn.) O. Ktze., Rev. Gen. Pl. 3 : 247 (1893)
 Sesamum sabulosum A. Chev., Étud. Fl. Afr. Cent. Franç. 1 : 229 (1913). Types :
 Chad, Baguirmi, *Chevalier* 9750, 9781 (P, syn.)

2. **S. indicum L.**, Sp. Pl., ed. 1, 634 (1753) ; P.O.A. C. : 365 (1895) ;
F.T.A. 4 (2) : 558 (1906) ; Z.A.E. 2 : 291 (1911) ; F.W.T.A. 2 : 243 (1931) ;
F.P.N.A. 2 : 253 (1947). Types : " India " in Herb. Sloane (BM, syn. !)
& no locality or collector (LINN, syn. !)

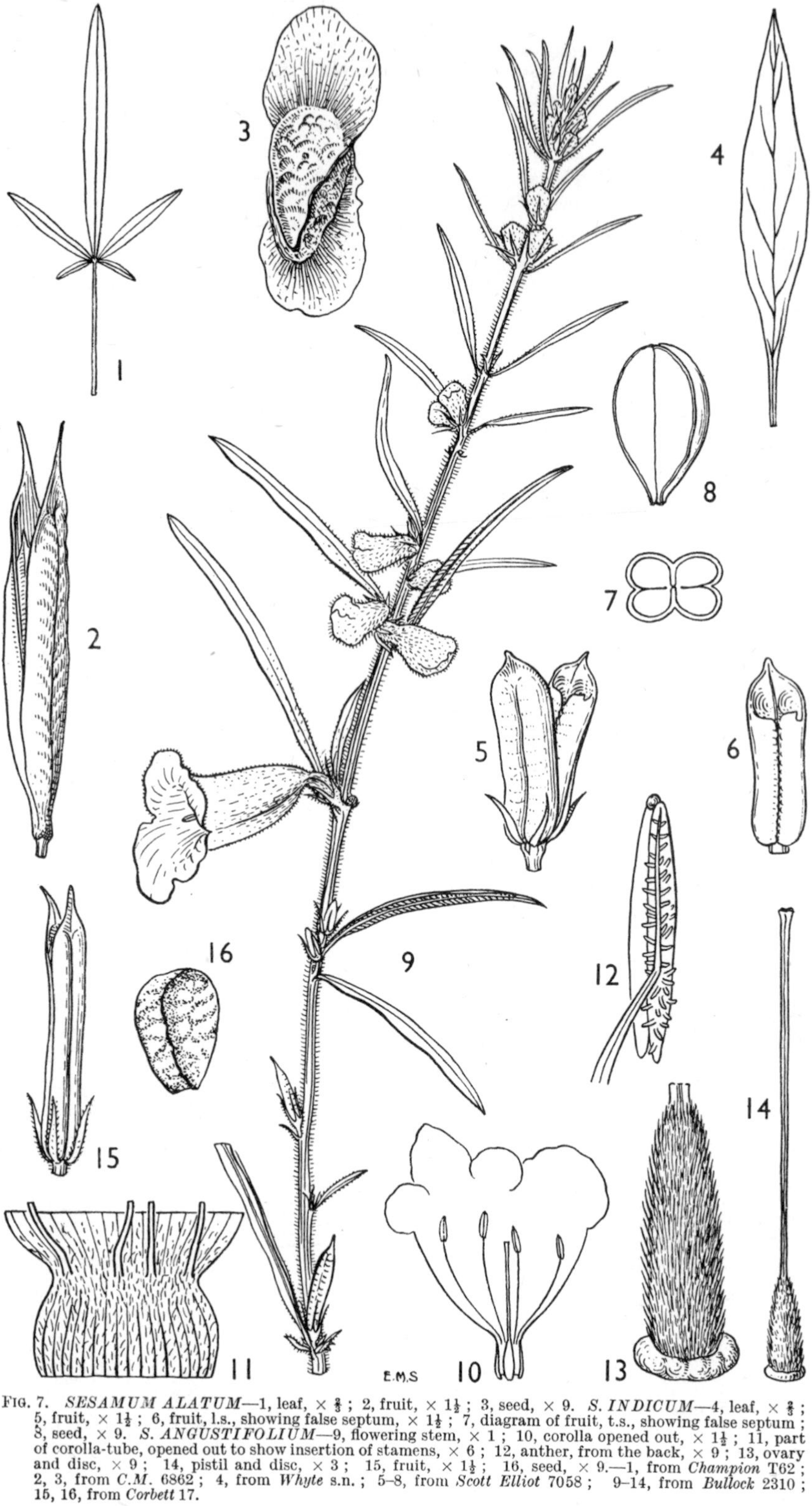

Fig. 7. *SESAMUM ALATUM*—1, leaf, × ⅔ ; 2, fruit, × 1½ ; 3, seed, × 9. *S. INDICUM*—4, leaf, × ⅔ ; 5, fruit, × 1½ ; 6, fruit, l.s., showing false septum, × 1½ ; 7, diagram of fruit, t.s., showing false septum ; 8, seed, × 9. *S. ANGUSTIFOLIUM*—9, flowering stem, × 1 ; 10, corolla opened out, × 1½ ; 11, part of corolla-tube, opened out to show insertion of stamens, × 6 ; 12, anther, from the back, × 9 ; 13, ovary and disc, × 9 ; 14, pistil and disc, × 3 ; 15, fruit, × 1½ ; 16, seed, × 9.—1, from *Champion* T62 ; 2, 3, from *C.M.* 6862 ; 4, from *Whyte* s.n. ; 5–8, from *Scott Elliot* 7058 ; 9–14, from *Bullock* 2310 ; 15, 16, from *Corbett* 17.

Erect annual herb, 10–120 (rarely 180) cm. high, simple or branched. Stems obtusely quadrangular, sulcate, finely pubescent to glabrescent, rarely pilose, usually more or less glandular. Leaves very variable, often heteromorphic, opposite or alternate ; lower leaves long-petiolate (3–11 cm.), ovate or ovate-lanceolate, 3-lobed, 3-partite or 3-foliolate, 4–20 cm. long, 2–10 cm. broad, rounded or obtuse at the base, acute at the apex ; margins often dentate; upper leaves more shortly petiolate (0·5–3 cm.), narrower, oblong-lanceolate to linear-lanceolate, 0·5–2·5 cm. broad, usually entire, and narrowly cuneate at the base (Fig. 7/4) ; all leaves thinly pubescent and more or less glandular, becoming glabrescent. Flowers white, pink or mauve-pink with darker markings. Calyx persistent; lobes oblong, pubescent, 2–5 mm. long. Corolla 1·5–3·3 cm. long. Filaments glabrous : anthers 2–3 mm. long. Ovary slightly compressed, 1–1·5 mm. long, more or less rounded at the apex, pilose. Capsule oblong-quadrangular, slightly compressed, deeply 4-grooved, rounded at the base and apex, then rather abruptly and shortly beaked at the apex, 1·5–3·2 cm. long (including the beak), 6–7 mm. broad, glandular and pubescent; beak 2–4 mm. or rarely shorter (Fig. 7/5–7). Seeds (Fig. 7/8) more or less horizontal in the capsule, not winged, 2·5–3 mm. long, about 1·5 mm. broad, black, brown or white, with faces smooth or rarely slightly veined, never rugulose or reticulate. Fig. 7/4–8.

UGANDA. No locality, Mackie Ethnological Expedition, *Roscoe* s.n. !
KENYA. Kilifi District : Kibarani, 26 Jan. 1945, *Jeffery* K.24 !
TANGANYIKA. Shinyanga, *Koritschoner* 2225 ! ; Rufiji District : Mafia Island, 14 Aug. 1937, *Greenway* 5094 !
ZANZIBAR. Pemba, Chonga-Chanjani, 14 Feb. 1929, *Greenway* 1442 !
DISTR. U? ; K3–5, 7 ; T2, 4, 6, 7 ; P. Cultivated in Europe, Asia, Africa and parts of Mexico and Cuba ; native of parts of India and Africa
HAB. Not indigenous in East Africa ; usually found in native gardens, grassland and by roadsides, on sandy soil ; 10–2500 m.

SYN. *Sesamum orientale* L. * Sp. Pl. ed. 1, 634 (1753). Type : Cult. in Hort. Cliff. (BM—Herb. Cliff., lecto. !)

NOTE. Cultivated for the seeds which are rich in oil.

3. **S. angustifolium** (*Oliv.*) *Engl.* in P.O.A. C. : 365 (1895) ; F.T.A. 4 (2) : 554 (1906) ; F.P.N.A. 2 : 253 (1947) ; W.F.K. 101 (1948). Type : Tanganyika, Unyamwezi, *Speke & Grant* (K, holo. !)

Erect or sometimes spreading, simple or branched herb 30–180 cm. high, stems subquadrangular, sulcate, pubescent to thinly pilose, becoming glabrescent. Leaves sessile, subsessile or lower ones sometimes shortly petiolate, narrowly linear to linear-lanceolate (rarely oblong-lanceolate), 2–12 cm. long, 0·1–1 cm. (some lower toothed leaves rarely up to 4 cm.) broad, margins entire or rarely undulate, sometimes the lower leaves coarsely and irregularly toothed, thinly pubescent to glabrous above, usually densely glandular below and thinly pubescent on the nerves becoming glabrous, cuneate at the base, acute, subacute or rarely rounded at the apex. Flowers pink, red, mauve or purple, often spotted within. Calyx persistent, lobes lanceolate to linear-lanceolate, acuminate, 5–9 mm. long, pubescent. Corolla 2–3·5 (rarely up to 4) cm. long, 0·9–1·8 (rarely up to 2·3) cm. diameter at the throat, pubescent or rarely glabrescent. Filaments arising from a band of hairs, anthers 3·5 mm. long, dorsifixed with papillae at the back. Ovary narrowly oblong-quadrangular, about 4 mm. long, densely adpressed-pubescent. Capsule narrowly oblong-quadrangular, deeply

* A. P. De Candolle in Pl. Rar. Jard. Genèv. 17 & 18 (1829), was the first author to reduce *S. orientale* to *S. indicum* making it a variety of this species. He says that in deciding to unite the two species he has chosen the name *indicum* in preference to *orientale* as this indicates the true country to which the species belongs.

4-grooved, rounded at the base, acuminate into a narrow beak at the apex, 1·2–2·4 cm. long, 3–4 mm. broad, thinly pubescent to pilose, beak 1–3·5 mm. long (Fig. 7/15). Seeds (Fig. 7/16) not winged, thick, about 1·5 mm. long, 1 mm. broad, faces and sides rugose. Fig. 7/9–16, p. 18.

Uganda.   Karamoja District : Lochoi, 24 May, *A. S. Thomas* 3527 ! ; Teso District : Serere, Nov.–Dec. 1931, *Chandler* 99 !

Kenya.   Kitui District : Kitui Hills, Jan. 1937, *Gardner* 3605 ! ; Tana River District : *Battiscombe* 258 !

Tanganyika.   Mwanza, April 1933, *Rounce* 289 ! ; Mpanda District : Ikuu, Jan. 1950, *Bullock* 2310 !

Zanzibar.   Zanzibar Is., unlocalized, *Toms* 99 !

Distr.   U1–4 ; K4–7 ; T1–8 ; Z;   Nigeria, Cameroons, Belgian Congo, Portuguese East Africa, Nyasaland, Northern and Southern Rhodesia, Angola, Bechuanaland, South West Africa

Hab.   In cultivated and waste areas, native gardens, roadsides and short grassland ; often cultivated as a vegetable, for its medicinal properties and for the oil in its seeds ; (15–) 600–2000 m.

Syn.   *Sesamum indicum* var. ? *angustifolium* Oliv. in Trans. Linn. Soc. 29 : 131 (1875) *S. baumii* Stapf in F.T.A. 4 (2) : 554 (1906). Type : Angola, Huila, between Kiteve & Humbe, *Baum* 959 (K, holo. !)

Variation.   The above species is very variable and includes both very erect and straggling forms.   The branched more spreading form has smaller flowers and shorter leaves and appears rather more woody.   A number of specimens, particularly those from the Lake Province (T1), have coarsely and irregularly toothed, sometimes lobulate, lower leaves, approaching *S. radiatum*.   If fruits are present on the specimens, they can be distinguished from those of *S. radiatum* by being narrower, less pilose and with longer beaks, whilst the seeds have thick sides like those of typical *S. angustifolium*.   If no fruits are available, an examination of the ovary will show that it is long and narrow and not so conspicuously white-pilose as in true *S. radiatum*. In my opinion *S. radiatum* does not occur in our area.   Again, there are specimens from Northern and Southern Rhodesia with blunter leaves and rather broader fruits than in typical *S. angustifolium*, these approach, and may possibly be *S. calycinum* Welw.   Except for the blunter rather broader leaves and rather shorter broader capsules there appears to be little difference between these two species.

4. **S. angolense** *Welw.*, Apont. Phyto-Geogr. 588 (1859) ; Trans. Linn. Soc. 27 : 51 (1869) ; F.T.A. 4 (2) : 555 (1906) ; P.O.A. C. : 365 (1895) ; W.F.K. 101 (1948), *? quoad descr. p.p., excl. fig.* 81.   Type : Angola, Pungo Andongo, *Welwitsch* 1645 (BM, holo. !, K, iso. !)

Simple or branched, decorative, densely leafy, shrubby perennial herbs, 0·8–3 m. high ; stems subquadrangular, sulcate, thinly pubescent.   Leaves subsessile or shortly petiolate, generally discolorous, narrowly oblong, oblong-lanceolate or rarely elliptic, 2–11 cm. long, 0·4–4 cm. broad, margins entire more or less inrolled and scabrid, upper surface glabrescent, lower surface white tomentose, more sparsely so on older leaves so that the glands are visible, cuneate at the base, truncate, rounded emarginate, subacute (rarely acute) and usually apiculate at the apex.   Flowers pink, red, purple, or pale mauve with deeper markings.   Calyx usually persistent ; lobes lanceolate or ovate-lanceolate, 5–10 mm. long, about 2 mm. broad at the base, acute or acuminate at the apex, pubescent.   Corolla large, 3·5–7 cm. long, 2–3 cm. in diameter at the throat, thinly to densely pubescent.   Filaments arising from a band of hairs near the base of the tube ; anthers narrowly oblong, 6 mm. long.   Ovary densely white-appressed-pilose, apex emarginate.   Capsule subquadrangular, 4-sulcate, 2–2·5 rarely 3 cm. long, 5–6 mm. broad, rather densely pubescent, becoming glabrescent, gradually narrowed into a flattened rather broad beak up to 5 mm. long.   Seeds not winged, about 2 mm. long, 1·5 mm. broad, faintly rugose on the sides and faces.

Uganda.   Masaka District : Buddu, Kyebe, Aug. 1945, *Purseglove* 1784 !

Kenya.   N. Kavirondo District : Kakamega (specimen cultivated at Nairobi), *Jex-Blake in C.M.* 6936 !

TANGANYIKA. Bukoba District : between Bukoba and Kyaka, Oct. 1940, *Tweedie* 543 ! ; Mpanda District : Ikuu, 22 Jan. 1950, *Bullock* 2311 ! ; Iringa District : Chiwanje, 18 Aug. 1933, *Greenway* 3575 !

DISTR. **U**4 ; **K**5 ; **T**1, 4, 6–8 ; Belgian Congo, Nyasaland, Northern and Southern Rhodesia, Angola

HAB. In wooded and upland grassland, by roadsides, in abandoned native cultivation, along river valleys and open woodlands, on black or red loam soil ; 350–2300 m.

SYN. *Sesamum macranthum* Oliv. in Trans. Linn. Soc. 29 : 131, t. 84 (1875). Type : Tanganyika, Usui, *Speke & Grant* (K, holo. !)
    S. macranthum Oliv. var. ? *angustifolium* Oliv. l.c. Type : Tanganyika, Uzaramo, Muhonyera, *Speke & Grant* (K, holo. !)

NOTE. This appears to be a fairly clear-cut species with large flowers and blunt leaves, which are white-tomentose on the lower surface. The species *S. calycinum*, mentioned under *S. angustifolium*, somewhat approaches *S. angolense* in its leaf-shape but differs in the smaller flowers and glabrescent or thinly pubescent leaves.
    The figure referred to this species in W.F.K. 101, t. 81, was in fact adapted from the Coryndon Museum painting of *Bally* 6071, which specimen is referred to *S. latifolium*.

5. **S. latifolium** *Gillett* in K.B. 1953 : 118 (1953). Type : Uganda, Karamoja District, Labwor, *A. S. Thomas* 3708 (K, holo. ! EA, iso. !)

Erect herb, 60–120 cm. high ; stems quadrangular, sulcate, more or less densely pubescent with brownish hairs. Leaves generally heteromorphic, lower ones large, long-petiolate (8–20 cm.) ovate-cordate or 3-lobed, 8–13 cm. long, 9–18 cm. broad, cordate at the base, upper leaves smaller, 0·5–5 cm. petiolate, ovate to ovate-lanceolate, 2–5 cm. long, 0·7–4 cm. broad, sub-truncate or cuneate at the base, inconspicuously to coarsely serrate on the margin, acuminate at the apex, pubescent on both sides more densely so below and on upper leaves ; mucilage glands present but not conspicuous. Flowers pale pink or pinkish mauve ; bracteoles conspicuous, pale brown, rather membranous, narrowly lanceolate, pubescent. Calyx ultimately deciduous ; lobes narrowly lanceolate, acuminate, rather similar to the bracteoles, about 5 mm. long. Corolla thinly pubescent and glandular outside, 2·5–2·8 cm. long, about 1 cm. broad at the throat. Filaments glabrous. Ovary adpressed-pubescent, narrowly oblong, about 6 mm. long, gradually narrowed into the style. Capsule shortly pubescent, oblong-quadrangular, 4-grooved, 2·7–3·1 cm. long (incl. beak), about 0·5 cm. broad, unilaterally gibbous at the base, rather abruptly narrowed into a short beak at the apex up to 5 mm. long. Seeds not winged, about 3 mm. long, 2 mm. broad, faces foveolate, sides deeply pitted, margins sharply angled.

UGANDA. Karamoja District : Labwor, 1300 m., 4 June 1940, *A. S. Thomas* 3708 ! & Iriri, near Napak, *Tweedie* 146 !

KENYA. Lake Rudolf, *Wellby* ! ; Northern Frontier Province : Tana River, Chebele, 28 Oct. 1945, *Adamson* 171 *in Bally* 6071 !

DISTR. **U**1 ; **K**1, 2 ; A.-E. Sudan, Ethiopia

HAB. Locally frequent among rocks in semi-desert scrub ; 300–1300 m.

NOTE. The material of this species is inadequate and it has been difficult to determine the specimens with certainty. The main variation is in leaf-shape and dentition. The type specimen shows lower leaves, which are coarsely serrate and subcordate, whereas *Adamson* 171 and *Padwa* 278 from the Sudan have only upper leaves which are longer, narrower and more cuneate at the base. Another Sudan specimen (*Myers* 9922) has some lower leaves which are broader and rounded at the base and link up with the type specimen.

INDEX TO PEDALIACEAE